Maria de Lourdes Ferreira Nunes
Maria Dayse Cardoso Pinto

An analysis of the use of media in biology classes at IFMA

Maria de Lourdes Ferreira Nunes
Maria Dayse Cardoso Pinto

An analysis of the use of media in biology classes at IFMA

How are biology teachers relating to the media throughout their teaching practice?

ScienciaScripts

Imprint

Any brand names and product names mentioned in this book are subject to trademark, brand or patent protection and are trademarks or registered trademarks of their respective holders. The use of brand names, product names, common names, trade names, product descriptions etc. even without a particular marking in this work is in no way to be construed to mean that such names may be regarded as unrestricted in respect of trademark and brand protection legislation and could thus be used by anyone.

Cover image: www.ingimage.com

This book is a translation from the original published under ISBN 978-3-330-99609-0.

Publisher:
Sciencia Scripts
is a trademark of
Dodo Books Indian Ocean Ltd. and OmniScriptum S.R.L publishing group

120 High Road, East Finchley, London, N2 9ED, United Kingdom
Str. Armeneasca 28/1, office 1, Chisinau MD-2012, Republic of Moldova, Europe
Managing Directors: Ieva Konstantinova, Victoria Ursu
info@omniscriptum.com

Printed at: see last page
ISBN: 978-620-8-58858-8

SUMMARY

CHAPTER 1	2
CHAPTER 2	5
CHAPTER 3	6
CHAPTER 4	28
CHAPTER 5	31
CHAPTER 6	40
CHAPTER 7	42
CHAPTER 8	46

CHAPTER 1

INTRODUCTION

Education is the basis of all human formation and organisation. The tools used throughout this process are extremely important for building a vision of the world, for forming participative and stimulated citizens. Starting from this point, there is a clear need for teaching/learning adjustments that meet these expectations, creating conditions that allow interconnections with the educational process and the evolution of technological resources as a means of achieving differentiated and meaningful learning. (MERCADO, 2007).

With the globalised and communicative world in mind, there have been constant changes in the history of didactics, in which there has been a need for resources to help as a tool to stimulate the educational process, becoming a differential in the development of classes and curricular activities.

Education needs to be rethought and alternative ways must be sought to increase teacher enthusiasm and student interest (FARIA, 2001). What role does technology play in this process of change? The application of technological resources in education is one that suggests changes in the pedagogical approach, leading subjects towards more creative activities and joint construction. Technological resources create a stimulating teaching and learning environment that provides opportunities for students to research and participate in the community with autonomy (MORAN, 2000).

Most educational institutions probably already have a range of technologies at their disposal. However, the presence of these tools does not necessarily mean that they are being used properly.

However, these tools can help the teaching-learning process in everyday life, but they end up abandoned, dusty in a room and becoming electronic rubbish.

In this context, it is worth emphasising that although education is equipped with new technological resources, it is still limited to the use of old technologies due to the lack of training aimed at including these tools. However, this allows the classroom to become a monotonous environment without any stimulus for the main elements of mobility in the process. It is up to the teacher to seek knowledge about the appropriate use of new technologies, since any resource used to mediate teacher/student interaction is a facilitator of learning (SCHWARZELMÚLLER, 2006).

Faced with this reality, using technology in favour of education means knowing how to use it as an auxiliary support in the search for quality in the educational process, given the obvious need for to provide a new vision in the teaching-learning process.

According to Melo (2010), "the use of technological resources, an instrument that contributes to interaction and mediation, as support for biology classes, is no longer an optional practice and has become a necessary resource for working with science content", because images, for example, as well as illustrating, contribute to the problematisation and contextualisation of such content.

In order to discuss the ways in which biology teachers use technological resources, this research was based on the following problem: What relationship do biology teachers at the IFMA - Monte Castelo Campus have with technological resources in their teaching practice?

Following this line of enquiry, the aim of the research was to investigate how biology teachers use technological resources in their teaching practice. Therefore, an analysis of the use of technological resources by IFMA teachers was carried out, seeking to analyse the application of technology in the educational environment, followed by field research to broaden the debate and finally contribute by giving greater visibility to recent changes and presenting a reflection on the future prospects and challenges that are being posed to the educational community.

CHAPTER 2

OBJECTIVES

2.1. General Objective

> To investigate how biology teachers relate to technological resources throughout their teaching practice.

2.2. Specific objectives

J Analyse contemporary bibliography on the subject in question;

J To survey the ways in which biology teachers are relating to technological resources in their classes;

J To identify the profile of the use of technological resources by biology teachers in state schools in São Luís - MA.

CHAPTER 3

THEORETICAL FRAMEWORK

3.1.. Teaching practice and technological innovations

The Jesuits were the first to exercise pedagogical practice in teaching in Brazil. The entire pedagogical act was determined by the "ratio studiorum", a pedagogical statute written by Ignatius of Loyola, containing rules that ranged from school organisation to pedagogical guidelines. The pedagogical ideas contained in the *ratio studiorum* gave rise to Traditional Pedagogy. The Jesuits expanded their pedagogy through music, theatre and dance, and made a major contribution to the arts. As Lima (2011, p.2) explains:

> The method of study contained in the Ratio comprised the trinomial study, repeat and dispute, prescribed in the rules of the Rector of the College, and as school exercises there was lecture, colouring lesson, composition and challenge, pedagogical practices that refer directly to medieval scholasticism, configuring itself as Traditional Pedagogy, which in its religious aspect, made education synonymous with catechesis and evangelisation.

In the Pombal period, pedagogical practice was geared towards the interests of the state. The royal classes were created, where each class was autonomous with a single teacher, and there was no interdisciplinarity between them. Teachers were poorly paid and declared owners of their classes for life . According to Saviani (2007, p.14):

> (...) with the expulsion of the Jesuits, pedagogical thought incorporated characteristics of Enlightenment ideas expressed in the enlightened despotism that led the Pombaline reforms public education. This orientation, which opened up space for secular pedagogical ideas, continued throughout the imperial period with the influence of spiritualist eclecticism, liberalism and positivism on pedagogical thought, becoming the vector of educational policy with the proclamation of the Republic. From then on, a secular state was established and, as a result, religious teaching was abolished from public schools.

Over time, many educators have joined the fight to reformulate education in Brazil. Until the approval of the National Education Guidelines and Bases Law, education in Brazil was dominated by the elite and mirrored social inequality. These transformations were based on a right acquired by the citizen, and so that teachers could apply a pedagogical practice that was consistent with their position on political and social issues.

In this context, it is important to note that pedagogical practice requires a meaningful direction for the human formation of individuals and processes that ensure the practical activity that corresponds to them. In other words, in order to make the educational process effective, it is necessary to give it an orientation, the aims and the means for its realisation, depending on the choices made in terms of the type of education you want and the type of society you aspire to.

It's not just today that technologies have entered education, they are part of a history that dates back to the military coup of 1964 with the technicist movement, but they didn't have the function of a resource, but of an instrument, aimed at reproduction and the division of labour based on Taylorist theory.

According to Druker (1993), today's society is undergoing profound changes characterised by a deep appreciation of information. In what is known as the Information Society, knowledge acquisition processes take on a prominent role and require a critical, creative professional with the ability to think, to learn how to learn, to work in groups and to know themselves as individuals (MASETTO, 1996). It is up to education to train this professional and, to this end, it is not just based on the instruction that the teacher gives the student, but on the construction of knowledge by the student and the development of new competences, such as: the ability to innovate, to create something new from what is known, adaptability to the new, creativity, autonomy and communication.

It is therefore clear that educators need to have a researcher profile; to formulate work proposals; to be able to put theory into practice and theorise practice, seeking to keep up to date, perfecting themselves in order to remain engaged with the system. However, they are not prepared for this, as many find it difficult to adopt technological innovations in their pedagogical practices without adequate training, which provides support for the teacher to be a researcher, to reflect on and plan their pedagogical actions with the help of various resources, including technological ones.

3.2. Teacher training in the face of technological resources

A question that often has to be asked is what kind of teachers are being trained? Who are they? What concerns and perspectives do they carry? Are they properly prepared for the new challenges posed by everyday reality? We can't lose sight of the reality of teaching in Brazil, which is heterogeneous, especially in terms of teacher training. In small and medium-sized towns, it is common to have lay teachers with no academic, pedagogical or, above all, scientific training, which makes it difficult to take bolder action in teaching practice. This situation seems to be one of the great challenges for society, which is to train teachers in sufficient quantity and quality to meet the diversity and needs of today.

Experiences of implementing technology in educational institutions have shown that teacher training is fundamental and requires a completely different approach (VALENTE*, et al,* 1997). Firstly, the implementation of information technology, which involves much more than providing teachers with knowledge about computers or methodologies on how to use the computer in their respective subject. There are other barriers that neither the teacher nor the administration can overcome without the help of experts in the field.

The digital inclusion of higher education teachers is a basic need within the developments and challenges posed by the Information Society. Not only teachers, but all people, regardless of age, gender or schooling, need to go through and remain in a continuous process of digital inclusion, as Silva et al (2005) point out:

> If the digital inclusion of teachers is an inherent need of this century, then this means that the citizen of the 21st century, among other things,

must consider this new factor of citizenship, which is digital inclusion. And that it is an ethical issue to offer this opportunity to everyone, i.e. the individual has the right to digital inclusion, and the person included has the duty to recognise that this right should be extended to everyone. In this way, digital inclusion is a process that should lead individuals to learn how to use ICT and access the information available on the networks. Especially information that will make a difference to their lives and to the community in which they live.

For the digital inclusion of teachers in higher education to become a reality, they need to undergo a new kind of literacy - digital literacy - in which they develop the skills and abilities needed to deal with the various teaching possibilities offered by Information and Communication Technologies (ICTs) (MERCADO, 1998).

In current experiences, teacher training to address this reality has not achieved its objectives, being restricted to initial, one-off, short-term training, and has been given little effective priority by universities. The solutions proposed are mainly part of continuing education programmes. Teachers working in virtual learning environments need to develop the competences and skills inherent in online education. According to Mercado (2007, p. 38), the basic competences for studying online should be encouraged and guaranteed to teachers right from their basic and initial training. These competences are:

> **Knowledge of online education resources** - introduction to the use of the Internet that may be useful from a didactic point of view, with an open attitude to change and experimentation, valuing technological resources and group work.

> **Analysing the possibilities and applying them within a didactic**

perspective - it is important to study the possibilities that resources offer for education, making them useful with benefits for the teaching-learning process.

> **Incorporating and integrating virtual learning environment interfaces into the teaching-learning process -** using interfaces to support teaching-learning processes, integrating them into the curriculum, favouring the modification of teaching methods. Exercise skills by planning curricular content that involves collaborative activities and Internet communication resources in class activities.

> **Expanding online spaces -** managing several spaces with different activities, several classrooms in different places, people, social and economic issues, through online activities;

> **Knowing how to interact with hypermedia content by associating** text, sound, photos, still and moving images and animation, allows students the freedom to navigate knowledge, to actively construct their learning, determining the paths to follow, according to their interests and at their own pace, on a journey of discovery, research and participation that will accompany them throughout their lives.

> **Knowing how to conduct an interaction in the virtual learning environment -** Development of the interactive processes that take place in the environment, from the perspective of co-operative work, involving collaboration in solidarity for the exchange of teachers' thinking and doing, working together in various instances and moments, taking place in groups and sub-groups, offering teachers the conditions to face the uncertainties and conflicts arising from current and future demands. A teacher must stimulate the students' "desire to learn", know how to work on interest and participation. In these mediation possibilities, collaborative construction is an option that can be used in online teaching.

The aim of this training, in addition to acquiring teaching methodologies, is to gain an in-depth understanding of the learning process, how it happens and how to intervene effectively in the teacher-student relationship and the virtual learning environment, providing students with favourable conditions for building knowledge.

3.3.. Integrating technological innovations into educational processes

Teaching resources are understood to be all the complementary instruments used by the teacher as an aid to the student's learning. However, among the resources that have accompanied systematised teaching in schools for a long time are: books, notebooks, written texts, blackboards and chalk.

Nowadays, new educational resources have been introduced, such as notebooks, data shows, tablets and digital whiteboards, which are considered to be resources linked to new technologies.

Faced with this reality, the application of technology has become highly relevant in various segments of society, especially in education, enabling innovation, research and the interaction of human knowledge. Through technological resources, information is produced and propagated rapidly throughout the world.

In the age of globalisation and digitalisation, a new attitude is required of man in every sense, especially in the way he thinks and acts. According to Rampazzo (2004, p.54):

> This new society, also known as the knowledge society, requires new skills and new attitudes, demanding an active, thinking, researching individual with intellectual autonomy. It is therefore up to the school, as the institution responsible for training the individual, to train people capable of dealing with technological advances. It needs to put learners in touch with the new communication and information technologies, as well as using technology in favour of education.

Taking these aspects into consideration, it should be noted that in contemporary times, characterised by the spread of technology, the revolution in communication tools and the evolution of information technology, new ways of living, relating, thinking and acting have emerged, giving a new face to society and consequently to individuals. Today's generations no longer have the same characteristics as those of the past. Techniques and technologies surround people's daily lives, implying new behaviours and a new way of thinking.

However, it is understood that the use of new technologies should not be reduced to a technicality that can be summarised as the use of new techniques and methodologies. New technologies, as we have seen, are the basis of a new society, so their application must be contextualised in a broader pedagogical project that takes account of this perception.

In this context, more precisely in education, the applicability of technological resources is proposed as a way of working towards achieving objectives. It is possible, through them, to work on playful situations and others that permeate the reality of academia associated with the use of information and communication technologies, relating them to scientific knowledge. The aim is to integrate the digital world with the content, making learning meaningful and directly affecting the student's social role.

It is therefore necessary to use technological resources as an educational tool, but teacher training is required. This training consists of understanding its limitations and possibilities, otherwise this tool will be underestimated or overestimated, causing misunderstandings for both the teacher and the student. According to Mercado (2000, p. 12):

> [...] educational institutions face the challenge not only of incorporating new technologies as teaching content, but also of recognising the conceptions that learners have about these technologies in order to design, develop and evaluate teaching practices that promote the development of a reflexive disposition towards technological knowledge and use.

The possible pedagogical properties that favour the challenge of studying, learning and teaching are therefore indisputable. With new educational tools, students develop autonomous learning skills, although preferably collectively, develop knowledge-building skills and motivate learning, leading students to analyse their use, their social impact and their implications.

It's important to emphasise how important it is to adapt new techniques to educational uses. This adaptation is not only in the equipment, but also in the people involved in the educational process, if we want to practise an active pedagogy, becoming ever smaller and more powerful. Orality and writing have taken on new forms of conception. Digital language incorporates orality and writing so that they can be used by means of techniques present in our contemporary times.

In view of Bellonni's arguments (2001, p.55) "The integration of technological innovations into educational processes will therefore depend on the concept of education for the new generations that underpins the sector's actions and policies". According to the above, the use of technology is not just restricted to new models of certain equipment and products, it changes behaviour.

It should be emphasised that the growing development of technology in the existing culture. It is clear that the transformation is not just in individual behaviour, but that of the entire social group. This means rethinking methods, reflecting on educational practice and understanding how education today can respond to the demands of society without becoming stuck in traditional models and practices.

According to Kenski (2007), a new type of technological society is emerging, determined mainly by advances in digital communication and information technologies and microelectronics. When these new technologies are disseminated socially, they change professional qualifications and the way people live, work, inform themselves and communicate with other people and the world.

Faced with this reality, technological resources bring a significant change to the teaching-learning process, as they make the student a participant in the educational process, as well as pushing the teacher to seek out new knowledge and adapt to the constant changes that society has undergone and that the school cannot remain indifferent to. According to Moran, 2000 apud SANTIAGO, 2006, p. 11:

> The new technologies make it possible to broaden the concept of the classroom, of space and time, to establish new bridges between the face-to-face and the virtual, between being together and being connected at a distance. However, the challenge in education is much greater than simply using technological resources; the important thing is to think about an education with technology that can be offered with quality.

It is therefore important to point out that in this perspective of teaching mediated by technological resources, the teacher moves from a school centred on knowledge, where the teacher has absolute mastery of what they are proposing, to a vision of a teacher who, when constructing knowledge together with their students, questions, doubts, faces conflicts, contradictions and divergences, enriching these actions with the support of technology.

It's also important to emphasise that technology is often brought to the teacher, but the teacher continues to use the same repetitive 'content transmission' methodologies which does not allow space for the student to create, learn, produce and become a citizen of the world. In short, teachers are not trained in the pedagogical use of technologies. Most of the time, these teaching professionals are more concerned with using the technologies they have at their disposal in order to cover the content, without worrying about the student, the one who needs to learn.

Technology as a means of facilitating learning is still something new in institutions, and it is "this new" that will open doors for knowledge to be seen as a new tool for achieving "educational know-how", because at times it is necessary to strip away traditionalist concepts and dress up in these new ideals. According to Borges (1999, p.136):

> Educational Informatics is characterised by the use of computers as support for the teacher, as an additional tool in the classroom, in which the teacher can use the resources made available to them. At this level, the computer is exploited by the specialist teacher to its full potential and capacity, making it possible to simulate, practise or experience situations, and even suggest abstract conjectures that are fundamental to understanding the knowledge or knowledge model that is being constructed.

Educational IT as a pedagogical tool undoubtedly helps in the process of building knowledge. It is a means and not an end, and should be used taking into account the development of curricular components. In this sense, the computer has become a powerful resource to support learning, with countless pedagogical possibilities, provided there is a reformulation of the curriculum, that new methodological and didactic models are created, and above all that the true meaning of learning is rethought, so that the computer does not become just another prop dressed up as modernity.

It's up to the teacher and student to make the most of all the resources that technology presents us with, in order to collaborate more and more with the acquisition of knowledge. It should also be emphasised that the student is first and foremost the end for whom the development of educational practices is applied, leading them to learn about and build their knowledge through interactivity with the learning environment.

It is up to the competent bodies, such as the Ministry of Education - MEC and the Ministry of Science, Technology and Innovation - MCTI, to democratise access to computers, promoting the socio-digital inclusion of our students. Leaders must also discuss and understand the pedagogical possibilities of this valuable resource. However, we need to be aware that it is not just the introduction of technology into the classroom that will bring about changes in students' learning; the computer is not a "panacea" for all educational problems.

Computer tools, especially the Internet, can be a resource rich in possibilities to help improve learning levels, provided that the curriculum is reformulated, new methodological models are created and the meaning of learning is rethought. Learning where there is room to promote the construction of knowledge. Knowledge, not as something that is received, but conceived as a relationship, or product of the relationship between the subject and their knowledge. Where the subject creatively discovers, constructs and modifies their own knowledge.

In this context, technological resources should be interconnected in the teaching-learning process, providing more creativity for educators and more meaning for students. Today's curriculum guidelines emphasise the importance of students learning to use mathematical procedures and concepts, as well as technological tools to solve problems. They also emphasise the importance of encouraging students to construct knowledge, cultivating self-esteem, respect for their colleagues' work and perseverance in the search for a solution. It also proposes that teaching be contextualised.

In this way, the technological resources being used by educators enable a better understanding of the content required in biology teaching, as well as allowing students to be an active part of the process and not just passive subjects of their own learning. This will also make them realise that learning is a continuous process in life.

Technologies in their most varied forms end up expanding the intellectual capacities of human beings, making available a range of information and access in different ways with different environments and tools and, along with all this evolution, students and educational institutions end up being affected and accompany these changes.

Thus, it is necessary to make it clear that today's educators must invest and be invested in training, theoretical and practical courses in order to learn how to handle all the tools available in education so that they feel secure in the tasks to be carried out and that this security is passed on constructively to the students.

3.4. Some types of technological resources used in the classroom

Traditional teaching methods such as the

blackboard/whiteboard now have to compete with computers, mobile phones, tablets, etc. According to Souza (2003), the incorporation of new technologies into education makes it possible to increase access to information for both the educator and the learner, and also expands the possibilities for interaction, collaboration and learner autonomy. It is not possible to ignore or avoid the changes in technology applicable to education, as it has been present in our daily routine through television, radio, the *internet* and information technology.

The classroom is no longer the same. Technology, once restricted to computer classes, is now part of the daily lives of students and teachers. The use of new technologies to mediate content is increasingly becoming a preponderant factor in the teaching and learning process.

In the society of knowledge and technology, it has become necessary to rethink the role of the school, more precisely, issues related to teaching and learning. According to Prado (2008), teaching that is organised in a fragmented way, which favours the memorisation of definitions and facts, as well as standardised solutions, does not meet the demands of this new paradigm.

According to Santiago (2006), Educational Technologies are all the resources that make it possible to enrich the art of teaching. Over time we can see that there has always been a search for ways to enrich the way we teach. Resources such as notebooks, pencils, blackboards and chalk are examples of technological advances incorporated into the teaching and learning process. In the same vein, other technologies such as television and video cassettes had their heyday a few years ago.

Today's educational institutions make use of the most varied forms of technology, working with computerised laboratories, educational software, the internet, video conferencing, platforms and many other tools that complement lessons.

Valente (2005) explains that the use of technological resources in the educational process makes it possible to explore different applications of knowledge and contributes to its construction. The author points out that certain practices and experiences with technology may not be in favour of this objective and it is up to the teacher to be aware and question whether or not the technology context is favouring the construction of knowledge. In this way, he points out that the teacher, based on their planning, should establish pedagogical objectives that will be enhanced by the use of technology resources in which they can integrate disciplinary content.

Among the most common digital technological resources used in the classroom , especially in higher education, are the following:

3.4.1 Internet

The Internet, as hypermedia, is seen as essential for the communication of teachers and students, because through it it is possible, at a cheaper and more favoured cost, to unite writing, speech and images with speed, flexibility and interaction, which was practically impossible a short time ago. There is also the viability of using the internet as a way of broadening knowledge.

3.4.2Notebook

As a result of the portability of the *notebook,* it is noteworthy that public school teachers are investing more and more time and money in their own training in technology in order to incorporate the use of computers and the internet in the classroom. Data shows that the number of teachers who bought a notebook (portable computer) with their own personal funds increased from 65 per cent to 76 per cent between 2010 and 2011 (G1, 2012).

3.4.3Data Show

The data-show is undoubtedly very useful, as this equipment gives quality to the lesson by projecting the synthesis of the points of the content to be taught in the projection of films, internet displays, visualisation of figures, sounds, images etc. It's a great tool for teaching practices. However, it must be emphasised that the misuse of this equipment in the classroom has caused enormous educational harm to students. The teacher is not a transmitter of information, a "repasser" of content, or anything similar. The relationship between teacher and student has to be humanised.

3.4.1 Digital whiteboard

The digital whiteboard is a human interface device used to

control the computer directly in the projection area, focussing the viewer's attention directly on the teacher or lecturer, as well as using didactic tools in its software that expand the possibilities of using an ordinary computer with various resources aimed at facilitating the presentation of multimedia information. The digital whiteboard is thus a large, touch-sensitive screen that allows you to post documents on the Internet, share files on the local network or send information by e-mail. The latest models have USB ports for transferring files to removable drives.

3.4.5Electronic Book or *eBook*

An electronic book, or eBook, is a digital version of a printed book that can be *downloaded* and read directly from a computer. There are various devices available on the market today for reading eBooks, such as *tablets and iPads.* Teachers should therefore guide their students towards reading *ebooks* or even lessons for further contextualisation.

3.4.6Sound Media

Sound media use sound to convey a message. In education, initiatives that use sound as a technological resource are still timid. But if we take a moment to look around us, we'll realise that technologies that use sound are part of people's daily lives: mobile phones, radios (car or portable), music players (MP3, etc).

There are various digital audio storage formats, the most common of which are wave, MPEG, RealAudio and MP3. The wave format is the standard used by Microsoft and IBM for audio storage. Unlike MP3, it doesn't compress files and takes up a lot of storage space (the files generated are large). MP3 is an audio compression format that has revolutionised the way sound files are made available. Files can be converted to MP3 with little loss of quality, reduced size (approximately 10 per cent of the storage space required for the digital audio CD format) and can be obtained over the Internet.

3.4.7Audio authoring software

To edit and create audio files, you can use authoring software specifically for this media. These are programmes available on the Internet that have tools for manipulating audio files. Here's a list of some available software: Adobe SoundBooth, Audacity, Gold Wave, Glame and Protux.

3.4.8Digital video formats

Nowadays you can edit videos and photos with musical backgrounds, narration and various effects on your computer. The important thing is to use video authoring software and some (basic) equipment such as a digital camera (mobile phone, camcorder, etc) and a microphone. It's important to highlight some of the software available for video editing: Microsoft Windows Movie Maker, Jing,

YouConvert, VirtualDub, among others.

3.4.9 *Pen-drive*

It's a device with flash memory and a USB connector. Its purpose is to store text files, spreadsheets, music and images, in a similar way to a floppy disc or CD. The advantages of this type of storage unit are its small size and ease of use. Excellent for storing lessons and slides to present to students.

3.4.10 . *Blog*

The term *blog* or weblog is a web page that is updated frequently through *"posts",* which are usually small in size and can also include audio, video, animations and other links. They can be used as a teaching resource , as an educational strategy, as a place to access information, as a digital portfolio, as a space for debate, etc. It's a worldwide success and its usability is growing by the minute, because it's easy to create, update and feed. The blog is an educational and communicative tool, a new channel for developing students' knowledge. It also promotes the organisation of one's own thinking, enabling interactivity between teacher and student, fostering and disseminating the teamwork carried out on the course.

It promotes building and living in an online community, consolidating the student through their participation on the web, providing students with: motivation, discovery, involvement, curiosity, among other benefits.

3.4.11 Mail and *Email*

Email is currently the most common way for people to exchange messages, files and communicate dynamically via the Internet. With an e-mail address you can send a message to anyone in the world. It's an asynchronous mode of communication, meaning that the recipient doesn't need to be connected to the computer when the message is sent, they will receive it.

With email, it is possible to organise a complete distance learning course, from enrolling students to receiving activities and making content available. A tool derived from email is the discussion list, which will be studied in the next instalment. Email can be used in a variety of ways: when writing texts: the teacher can follow the construction of a text via email. All versions of the texts (with the appropriate corrections) are stored and can be used as a history. The teacher can email all the students a text to read and analyse beforehand. Based on this text, a discussion is proposed on one (or more) topic(s) based on the text. In the case of distance learning, a discussion forum or chat can be used as an interactive tool for discussing the text.

In addition, building and experimenting with new models for producing and transmitting knowledge is essential if we are to meet the challenges of this new era. To this end, emerging practices in the digital world are good examples. The free software movement, whose fundamental characteristics are collaboration and networking, is presented by Nelson Pretto (2008) as a stimulus for the introduction of a collaborative logic that is essential to education.

CHAPTER 4

METHODOLOGY

4.1 Characterisation of the study site

This research was carried out at the Federal Institute of Education, Science and Technology of Maranhão, Monte Castelo campus, located at Avenida Getúlio Vargas, number 4, in the Monte Castelo neighbourhood, São Luís/MA. The Federal Institute of Maranhão (IFMA), based in São Luís, was created by Law No. 11.892, of 29 December 2008, through the integration of the Federal Technological Education Centre of Maranhão. It is an Autarchy operating in the state of Maranhão, with administrative, patrimonial, financial, didactic-pedagogical and disciplinary autonomy. It should also be noted that the degree courses, as well as special pedagogical training programmes, are aimed at training teachers for basic education, especially in the areas of science and mathematics, and for professional education.

The Biology course at IFMA Monte Castelo runs in the evening in the regular mode and the Biology course linked to the National Teacher Training Plan (PARFOR), under the general coordination of Prof. MSc Eliane Maria Pinto Pedrosa, runs at weekends under the coordination of Prof. Yrla Nívea Oliveira Pereira.

The course department is headed by Prof Dr Luis Fernando Gomes and the coordinator Prof Dr Naiza Maria Nogueira. The department has a total of 09 (nine) teachers, but 01 (one) is on leave for a training programme.

4.2 Type of Study

As this research deals with the relationship between teachers and teaching practices geared towards the use of technological resources during lessons, it is considered to be a qualitative and quantitative research approach. In order to better define the purposes of this research, it was classified according to Vergara (2003). In this sense, in terms of purposes, this research is descriptive and explanatory. In terms of means, it is bibliographical and field research.

It is descriptive because it exposes the teaching practices of Biology teachers at the Federal Institute of Education, Science and Technology of Maranhão - IFMA. It is explanatory because the actions studied are organised into data that is easy to understand, justifying and explaining their main reasons and the "why" of things. In other words, it tries to identify the factors that lead to the use or not of technologies in biology classes.

In terms of means, it is bibliographical, since books, scientific articles and electronic documents on the subject were used. Fieldwork, as the teachers on the Biology course at the IFMA - Monte

Castelo Campus were analysed. In order to achieve the proposed objectives, it was necessary to establish direct contact with the teachers within their reality, in an attempt to understand their perceptions by analysing and studying their speeches.

4.3 Data collection

This research involved all the Biology teachers who were working in full time, which corresponds to a total of eight teachers. It should be emphasised that the research focused exclusively on teachers who teach the specific subjects included in the undergraduate curriculum.

Teachers were interviewed using a questionnaire (attached) containing eight closed questions (multiple choice).

CHAPTER 5

RESULTS AND DISCUSSIONS

5.1. Professional identification of IFMA biology teachers

The survey shows a very diverse range of working time at IFMA, with only 25% of teachers having been at the institution for more than ten years, 50% between 7-10 years, 12.5% between 1-2 years and 12.5% between 4-7 years (Figure 1). This data shows that the majority of teachers on the Biology course at the IFMA Monte Castelo Campus have been working at the institution for a significant period of time. Institutional ties are an element that should be taken into account, as they help teachers to develop long-term planning and thus better understand the educational process in which they are involved.

Figura 1: Years of activity by IFMA biology teachers.

5.2 Biology teachers' time in the teaching profession

According to the data collected on the length of time teachers have worked in the teaching profession, it was observed that 86% of teachers have been working for more than 10 years, and 14% have only worked for between 4 and 7 years, as shown in Figure 2.

Experience in all professional careers is of the utmost importance, but in teaching it is highly prioritised. It is not something that is easily acquired. Experience is something that can only be acquired by living and doing. The data shows that IFMA's biology teachers have experience in pedagogical practice, as 86% have been teaching for a decade or more, so it is to be expected that they have a solid academic background and multidisciplinary knowledge. Therefore, work experience is just as important for learning as the theory taught in the classroom. Paulo Freire once said that "it is necessary to work with the reality of the student's daily life for effective learning. It is learning by doing, with learning by thinking about what is being done". According to Brito (2006), "the teacher's thinking is therefore built on their individual experiences and on the exchanges and interactions with their peers".

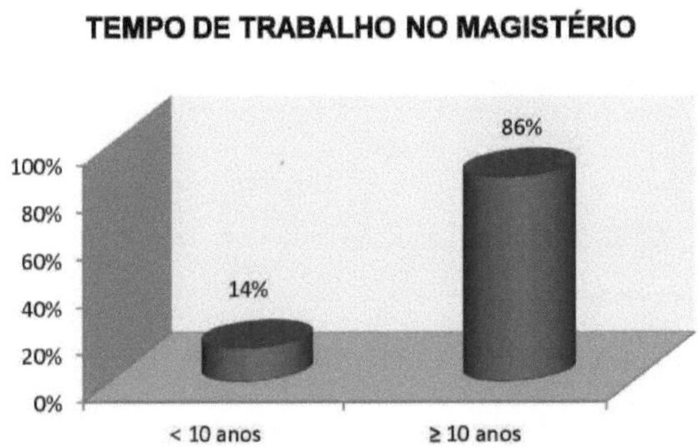

Figura 2: Years of teaching experience of IFMA biology teachers

5.3 Biology teachers' working conditions

According to the item on the work regime, 75% have exclusive dedication at IFMA, 12.5% work 40 hours and 12.5% work 20 hours (Figure 3). The vast majority today are DE teachers, and in many cases 20-hour teachers. Based on the data, the institute has a professional staff committed to education and the institution, which favours the quality of teaching. If teachers have more time to develop their work, we can deduce that they will increase their output as educators and researchers. And in the end, the winners are the students, the teachers, society and, above all, the institution.

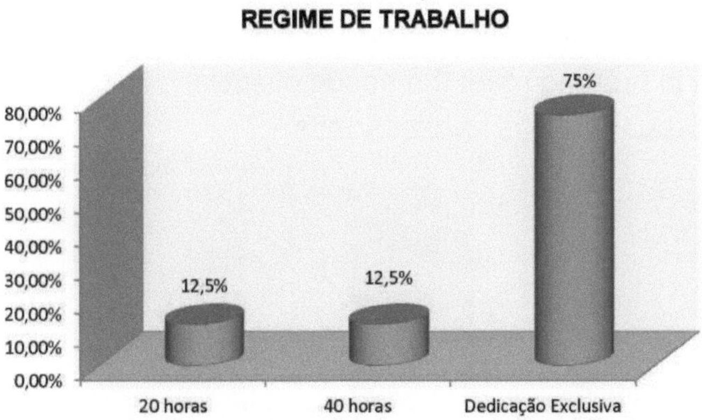

Figura 3: IFMA biology teachers' working conditions

5.4 Biology teachers' undergraduate programmes

In relation to the degree course and qualification, it was noted that 25% only have a degree, and 75% have a degree and a bachelor's degree in Biology, showing a positive point in the quality of teaching at the Institute in relation to Biology teachers, which we can see in Figure 4.

Pedagogical training is the construction of the teacher's initial training process, as the graduate acquires knowledge that makes up their reference frameworks for the profession. Lopes (2010) states that "although initial training alone cannot train teachers to work in an environment marked by unpredictability and constant change, it can provide a solid foundation for future learning". The graduate teacher has specific education subjects that are important for teaching practice, while the bachelor's degree trains the professional researcher (NUNES, 2001). Therefore, the data shows that the Institute has the majority of educators trained in both areas, in which they are trained to develop studies and stimulate research with innovation, with the aim of training critical, ethical, responsible citizens, with an entrepreneurial vision, capable of developing sustainable actions in order to meet the needs of society.

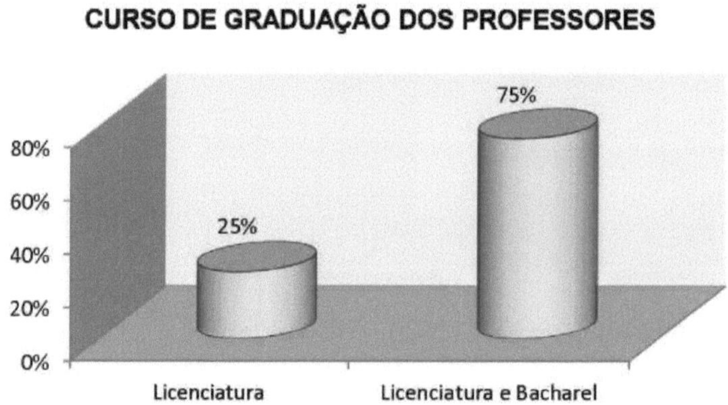

Figura 4: Graduation course for IFMA biology teachers.

5.5 Qualifications of IFMA biology teachers

In terms of qualifications, 37.5% have a Master's degree, 50% have a PhD and 12.5% have a specialisation (Figure 5). In view of the above, it can be seen that the Institute has a qualified teaching staff, complying with Law 9.394/96, which requires at least one third of the teaching staff to have a master's or doctorate degree.

However, we must not forget that a good teacher is not necessarily the one with the highest academic qualification, but the teacher who has experience and knowledge of a particular subject, which will contribute to the students' learning.

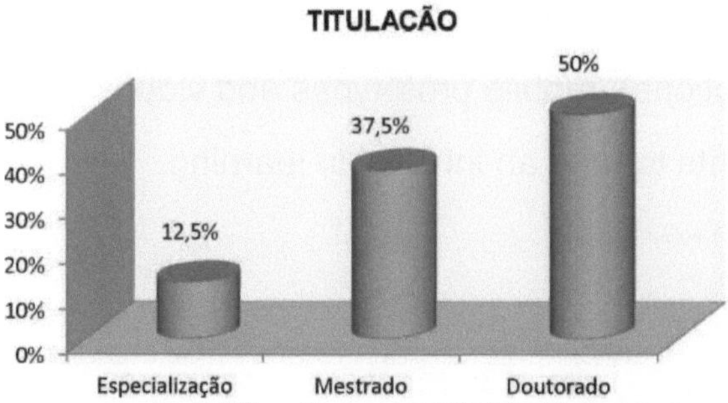

Figura 5: Qualifications of IFMA biology teachers.

According to the identification data of IFMA's biology educators, it was possible to see that all of them have adequate qualifications and significant experience to carry out their activities with mastery and quality, which is reflected in the quality of teaching and learning.

5.6 The use of technological resources in biology teaching by IFMA teachers

Based on the data collected on the use of technological resources in biology teaching, it was found that 100% of teachers frequently use data shows and computers. According to Fernandes (1998), slides allow for high-resolution projection, emphasising colour, beauty and detail, visible from anywhere in the classroom.

50% use the internet, 30% educational software, 20% social networks, 10% use the digital whiteboard, Moodle Platform and others (Figure 6). Therefore, the use of technological resources shows how significant student learning can be in the subject of Biology. As we have seen, classes should be diversified with the use of data shows (slide projections), mobile prototypes and videos. The aim is to encourage students to take an interest in learning.

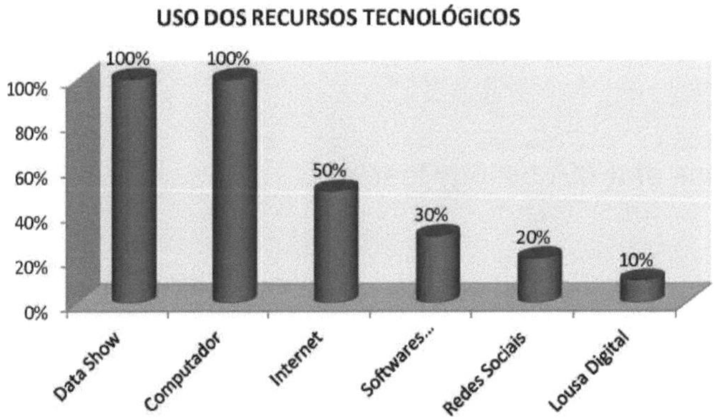

Figura 6: Use of technological resources by IFMA biology teachers

In this context, it is assumed that all of IFMA's biology teachers are trained and pass on the content through the resources with ease and mastery, providing students with an era of technological advances, in which their learning takes place more effectively, awakening them to a new vision of learning.

5.7 Reasons why IFMA biology teachers use technological resources

As for the reasons why teachers use technological resources in the classroom, 100% say that they save time, while 90% of teachers say that they make their work easier, 50% use them because they modernise the classroom and 30% believe that they make the classroom more attractive (Figure 7). So, with the teachers' effort and desire to provide quality education, and the government's investment in updating these professionals, a new pedagogical practice is gradually introduced into the classroom and technology narrows this communication channel between technology and teaching. It's therefore an intelligent way of being ahead in the generation of knowledge, transforming the way we educate and evolving the pedagogical process of teaching and learning.

The reasons for using these resources point to tools that complement the usual teaching activities and which, due to their flexibility and scope, favour the creation of knowledge networks (CUNHA, 2012). However, given the diversity of new pedagogical resources that are being made available, it is necessary for teachers to receive adequate training to work with these new pedagogical resources, being encouraged to work with this new practice, so that the educational process can really be renewed and improved and only then can the greater objective be achieved, which is to serve as a facilitator of the teaching/learning process, leading teachers to exercise their role as mediators in the construction of their students' knowledge.

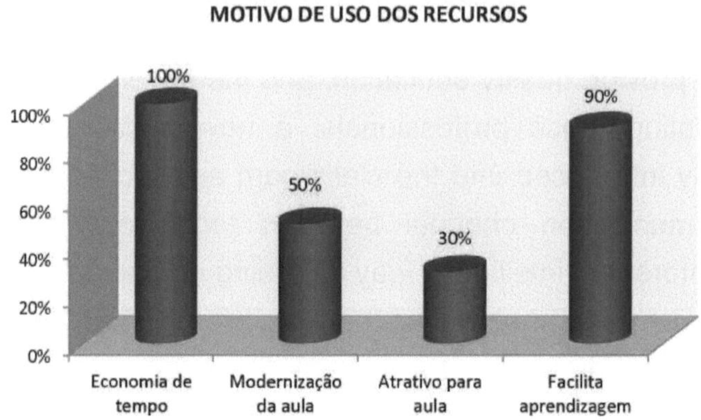

Figura 7: Reason for use of technological resources by IFMA biology teachers

Although all the teachers use technological resources, they still encounter problems related to the internet and machine maintenance.

7.8. IFMA biology teachers' opinions on the use of technological resources

Regarding the use of technological resources, 100% of the teachers said that these resources function as facilitators of teaching work. 87.5% of teachers believe that they can mediate content and learning and 37.5% see new technologies in teaching as a fad derived from the globalisation phase (Figure 8). Therefore, the traditionalism rooted in education is increasingly being remedied with the help of technological resources used by teachers in their classes. And technology-mediated teaching has modernised learning through changes to an expository model of content (SCHWARZELMÚLLER, 2006).

Figura 8: IFMA biology teachers' opinions on the use of technological resources

CHAPTER 6

CONCLUSION

In this context, it can be concluded that the teachers of the IFMA Biology course understood the need to use technological resources because they favour the dynamism of their classes and contribute to teaching and learning, exercising their role as mediators in the construction of their students' knowledge.

In terms of qualifications, it was realised that the institute has a staff of teachers who are qualified according to their area and who teach content in a creative way, seeking to build knowledge collectively through technological resources, showing that the use of these resources is already taking place effectively at the institution.

The analysis showed that all the biology teachers at IFMA have a close relationship with the use of technological resources, transforming the way they educate and evolving the pedagogical process of teaching.

The use of technological resources in education is a new process that aims to help and improve the teaching-learning process. However, even if IFMA's biology teachers have mastered the use of these resources, they need to look for new alternatives to obtain related knowledge, since technology is advancing at a high speed.

The use of technological resources, a tool that contributes to interaction and mediation, to support biology classes has ceased to be an optional practice and has become a necessary resource for working with content and images, helping to problematise and contextualise this content. In this way, as they are easily accessible resources that help the teaching-learning process, they can and should be used pedagogically and didactically in disciplinary and interdisciplinary activities.

However, in the face of the diversity of new pedagogical technological resources that are being made available, it is necessary for teachers to receive adequate training to work with these new resources, being stimulated to work by providing an improvement and renewal in the educational process.

Therefore, in order to see possibilities for success in the use of new technologies in education, it is necessary to consider teachers not just as the executors of an already defined computerisation project. It is therefore up to higher education teachers to be aware of and adapt to the technologies that are now being introduced, because the knowledge society now demands a new professional profile.

CHAPTER 7

REFERENCES

BELLONI, Maria Luiza. **Educação à distância.** 2ª edição. Campinas, SP: Autores Associados, p. 55, 2001.

BORGES NETO, H. Uma classificação sobre a utilização do computador pela escola. **Revista Educação em Debate,** ano 21, v. 1, n. 27, Fortaleza, 1999.
07/2011-03-23_Re-_BloggLinkUp_Request.jpg> Accessed 15 Jan. 2014.

BRAZIL. Ministry of Education. National Education Guidelines and Bases Law, No. 9.394/96. Brasília: MEC, 1996.

BRITO, A. E. Teacher training: rethinking work and teaching knowledge. In: MENDES SOBRINHO, J. A. de C.; CARVALHO, M. A. de (Org.). Formação de professores e práticas docentes: olhares contemporâneos. Belo Horizonte: Autêntica, 2006.

CONCEITOS, de. Available at:< http://conceito.de/recursos-tecnologicos>. Accessed on: 10 January 2014.

CUNHA, R. M. R. et al. Technological Resources as Potentialisers of Interdisciplinarity in the School Space. São Paulo: Unindu, 2012. Available at: . Accessed on: 08/06/2014.Available at:< http://www.bdtechie.com/ wpcontent/uploads/2011/

DRUCKER, Peter F. **Post-Capitalist Society.** São Paulo : Pioneira, 1993.

FARIA, E.T. The teacher and new technologies. In: ENRICONE, D.(org). Being a Teacher. Porto Alegre, Edipucrs, p.57-72, 2001.

FERNANDES, H. L. A naturalist in the classroom. Ciência & Ensino. Campinas, Vol. 5, 1998.

KENSKI, V. M. **Education and technologies:** The new rhythm of information. Campinas: Papirus, 2007.

LIMA, Daniela F. C. F. Ratio studiorum: pedagogical method of the Jesuits. 5th Postgraduate Congress. Available at:http://www.unimep.br/phpg/ mostraacademica/anais/5mostra/5/31.pdf Accessed on: 23 Dec. 2013.

LOPES, R. P. From graduation to the classroom: the process of learning to teach in different times and spaces. **Educar em Revista,** Curitiba, 36, p. 163-179. 2010.

MASETTO, M.; ALONSO, M. **Formar educadores para um mundo em transformação.** São Paulo: PUC/SP, 1996.

MELO, J. F. R. **Development of Practical Experimental Activities in Biology Teaching - a case study.** Postgraduate programme in science teaching professional master's degree in science teaching, UNIVERSIDADE BRASÍLIA: p.12, 2010.

MERCADO, Luís P; SILVA, Maria L. Use of virtual learning environments in teacher training. In: MERCADO, Luís P. (org). Pathways in teacher training with information and communication technologies in education. Maceió: Edufal, p. 175-196, 2007.

MERCADO, Luís Paulo Leopoldo. Teacher training and new technologies. In **New technologies in education:** reflections on practice. Maceió: Edufal, p.12, 2000.

MERCADO, Luís Paulo Leopoldo. **"Teacher training and new technologies. "***Ibero-American* Congress on *Information Technology in Education, IV.* 1998.

MORAN, José Manuel (et al). **New Technologies and Pedagogical Mediation.** 12ªed. Campinas: Papirus, 2000.

NUNES, Célia Maria Fernandes. Teaching Knowledge and Teacher Training - A Brief Overview of Brazilian Research. **Revista**

Educação e Sociedade, ano XXII, n° 74, Abril (2001).

FREIRE, Paulo. Pedagogy of autonomy: knowledge necessary for educational practice. 3rd ed. Rio de Janeiro: **Paz e Terra,** 148 p. 2004.

PRETTO, N. DE L. Digital culture and education: networks now! In: PRETTO, N. L.; SILVEIRA, S. A. (Org.). **Além das redes de colaboração:** internet, diversidade cultural e tecnologias do poder. Salvador, Edufba, 2008.

RAMPAZZO, Sandra R. dos Reis; RAMOS, Corina; VALENTE, Silza M. Pazello. Teacher training: pioneering distance learning experiences in the Brazilian context. **UNOPAR Científica:** ciências humanas e educação, Londrina, v. 5, p. 54, 2004.

SANTIAGO, Dalva Gonzalez. **New Technologies and Higher Education: rethinking teacher training.** Master's dissertation in Education. PUC- Campinas: 2006.

SAVIANI, Derme vai. **History of pedagogical ideas in Brazil.** Campinas: Autores Associados, p. 14, 2007.

SCHWARZELMULLER, Anna F.; ORNELLAS, Bárbara. Digital Objects and their Use in the Teaching-Learning Process. In: **1st Latin American Conference on Learning Objects.** 2006.

SILVA, Helena et alli. Digital inclusion and education for information competence: a question of ethics and citizenship. Brasília: **Ciência da**
Informação, v.34, n.1, p. 28-36, jan./abr.2005.

SOUZA, Carlos Henrique Medeiros. **Communication, Education and New Technologies.** Campos dos Goytacazes, RJ: Editora FAFIC, 2003.

VALENTE, J.A. Research, communication and learning with computers. The role of the computer in the teaching-learning process. In: BRAZIL, Ministry of Education. **Integration of**

Technologies in Education. Distance Education Secretariat. Brasília: Ministry of Education, Seed, 2005.

VALENTE, JA & Almeida, F.J.). Analysing Informatics in Education: the question of teacher training. **Revista Brasileira de Informática na Educação,** Sociedade Brasileira de Informática na Educação, n° 1, p. 45-60,1997.

VERGARA, Sylvia Constant. **Research Projects and Reports in Administration.** 4. ed. São Paulo: Atlas, 2003.

CHAPTER 8

ANNEX

FEDERAL INSTITUTE OF EDUCATION, SCIENCE AND TECHNOLOGY OF MARANHÃO DIRECTORATE OF HIGHER EDUCATION DEPARTMENT OF HUMAN AND SOCIAL SCIENCES TITLE: TEACHERS AND THE USE OF TECHNOLOGICAL RESOURCES: AN ANALYSIS OF THE BIOLOGY COURSE AT THE FEDERAL INSTITUTE OF EDUCATION, SCIENCE AND TECHNOLOGY OF MARANHÃO - MONTE CASTELO CAMPUS

QUESTIONNAIRE

Dear teacher,

The purpose of this questionnaire is to collect data on the use of technological resources by IFMA teachers and to contribute to the construction of new pedagogical mediation spaces. The data collected will be used exclusively for academic purposes within the scope of the monograph named above. Please feel free to complete the questionnaire and be as truthful as possible. Participation in the research is voluntary. Given the importance of confidentiality, you should not record your name on the questionnaire.

- PROFESSIONAL IDENTIFICATION DATA

1. How long have you worked at IFMA?

( )1 and 3 years ( )4 and 7 years ( ) Between 7 and 10 years ( ) More than 10 years

2. How long have you worked as a teacher?

( )1e3years ( )4e7years( ) Between 7 and 10 years ( ) More than 10 years

3. How do you work?

( ) 20 hours

( ) 40 hours

( ) Exclusive dedication

4. What is your degree and qualification?

Course ____________________________

( ) Licence ( ) Bachelor

Titling

( ) Specialisation

( ) Master's degree

( ) Doctorate

- INFORMATION ON THE USE OF TECHNOLOGICAL RESOURCES IN BIOLOGY TEACHING

5. Which of the courses listed below do you use in your classes?

( ) Data show

( ) Computer

( ) Overhead projector

( ) Internet

( ) Social networks

( ) Educational software

( ) Teleconference

( ) Videoconferencing

( ) Digital whiteboard

( ) Moodle Platform

( ) Chat

( ) Other____

( ) I don't use any technological resources in my classes (If you tick this box, please justify it)

6. If you have ticked some of the boxes in question 5, what are the criteria that lead you to use them?

( ) Saving time

( ) Facilitate learning

( ) Optimise content presentation

( ) Ecological stance (content economy)

( ) In tune with the digital age

( ) Modernising the classroom

( ) Insertion of a playful element in the presentation of content

( ) Use of the virtual environment as an attraction for teaching content

7. If none of the resources listed in question 5 are part of your classroom routine, what are the reasons you don't use them?

( ) Technological resources do not contribute to the construction of knowledge.

( ) The school I work at has no technological resources for teachers to use in the classroom.

( ) My students have difficulty accessing the internet.

( ) I prefer to teach by making small illustrations/drawings on the board.

( ) I don't know the didactic-pedagogical potential of technological resources.

( ) Other reason.

8. What is your opinion on the use of technological resources in the classroom by teachers?

( ) Lack of competence to teach the lesson in another way;

( ) A fad imposed by the consumerist wave of society;

( ) Possibility of mediating content and learning;

( ) A trend in the current phrase of economic globalisation;

( ) Resource that facilitates teaching work;

( ) Other opinion

THANK YOU!

MIX
Papier aus verantwortungsvollen Quellen
Paper from responsible sources
FSC® C105338

Printed by Books on Demand GmbH, Norderstedt / Germany